U-BOATS IN THE ATLANTIC

An ocean-going Type IXB with a characteristic broad beam comes into Lorient in May 1941. *U-109* was commanded by Kapitan-Leutnant Fischer and was lost two years after this photograph was taken (86MW/4254/30).

PAUL BEAVER

U-BOATS IN THE ATLANTIC

A selection of German wartime photographs from the Bundesarchiv, Koblenz

PSL Patrick Stephens, Cambridge

First published in 1979

British Library Cataloguing in Publication Data

U-boats in the Atlantic. – (World War 2 photo album; 11).
1. World War, 1939-45 – Naval operations, German – Pictorial works
2. World War, 1939-45 – Naval operations – Submarine – Pictorial works
I. Beaver, Paul II. Series
940.54'51 D781

ISBN 0 85059 386 7 (casebound)
ISBN 0 85059 388 3 (softbound)

Photoset in 10 pt Plantin Roman. Printed in Great Britain on 100 gsm Pedigree coated cartridge and bound by The Garden City Press Limited, Letchworth, Hertfordshire, SG6 1JS, for the publishers, Patrick Stephens Limited, Bar Hill, Cambridge, CB3 8EL, England.

CONTENTS

Acknowledgements
The author and publisher would like to express their sincere thanks to Mrs Marianne Loenartz of the Bundesarchiv for her assistance, without which this book would have been impossible.

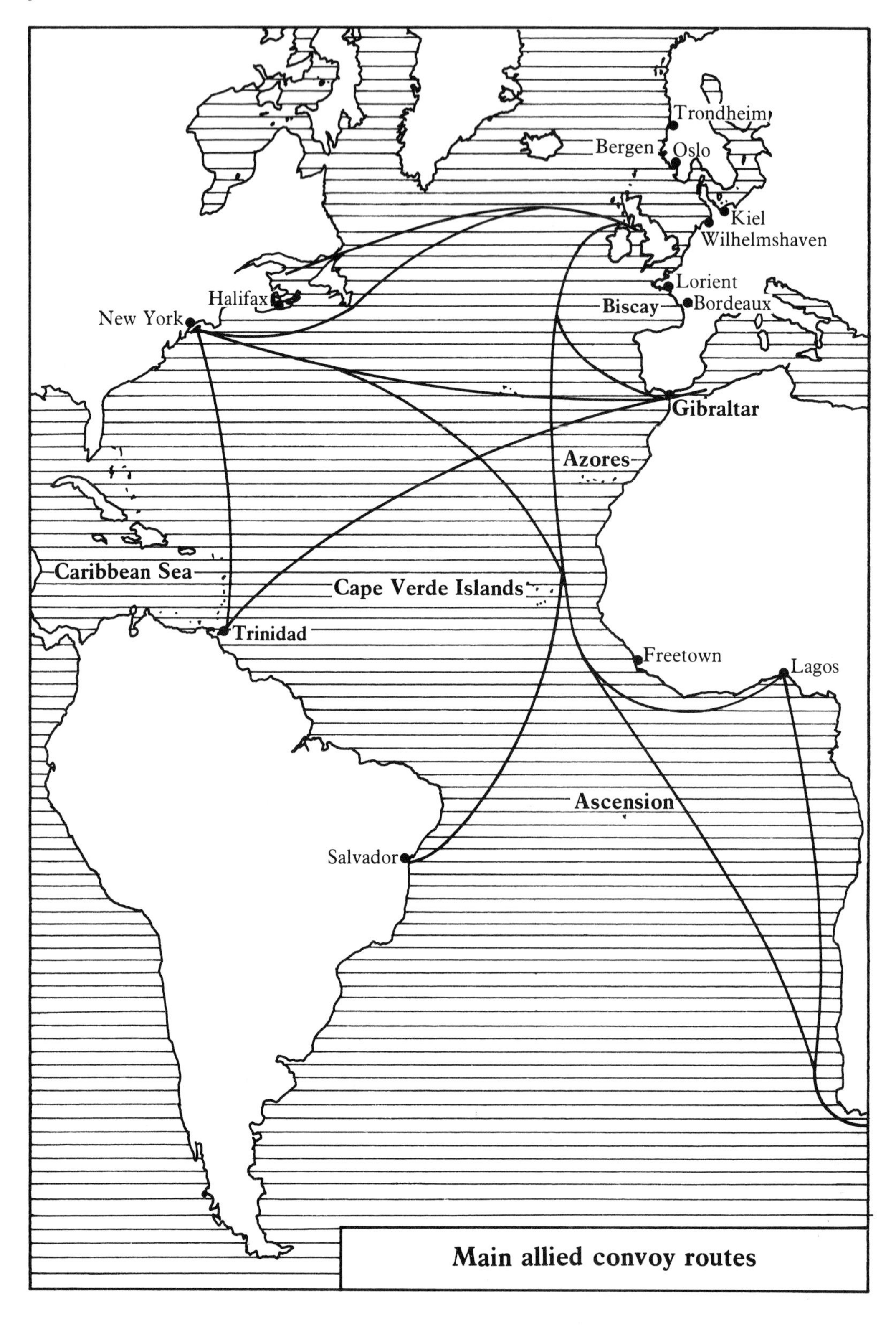

Main allied convoy routes

AUTHOR'S INTRODUCTION

This account, *U-boats in the Atlantic*, gives an insight into some aspects of the U-boat war in what was probably the most important 'battle' in World War 2.

Churchill said: 'If we [the British Isles] are cut off, if we lose the war at sea, nothing else will count.' It nearly happened.

The U-boat Fleet officially began to be revitalised in 1935 and, by 1939, the German Kriegsmarine was level pegging in numbers of operational submarines with the Royal Navy, having 56 boats in commission. Konter-Admiral Dönitz (later Gross-Admiral Dönitz who became the second Führer of Germany in 1945) knew, however, that even with the all-out effort of Admiral Raeder's Z-Plan, he would only have 250 submarines available by 1944. Had he had 300 units in 1939 he could have successfully strangled Britain's raw material life-line – a point worth remembering in the modern context. It is a sobering thought to note that, in all, 1,170 U-boats numbering U–1 to U–4712 were eventually built in only ten years.

The Z-Plan of Nazi Germany was drawn up in 1938 for the ten years until 1948. This ambitious scheme included the building of the following warships:

Aircraft carriers	4
Battlecruisers/Battleships	10
Pocket battleships/Heavy cruisers	20
Light cruisers	48
Scout cruisers	22
Destroyers	68
Torpedo boats	90
Submarines	250
Minesweepers/Minelayers	10
Auxiliary vessels	328
Total	**850**

On September 3 1939, 39 submarines were waiting for action in the North Sea and the Eastern Atlantic, hoping to cut off the British- and Empire-registered merchant ships which were scurrying for a safe haven – 102 being sunk. This was the time of the individual U-boat commands when it was the flair of the 'Aces' which counted rather than the combined firepower of the Wolf Packs of the mid-Atlantic air gap.

One of the first warship victims of the war, an event which shocked the whole nation, was HMS *Royal Oak*. This 29,000-ton battleship was caught sleeping in the 'safe' naval anchorage of Scapa Flow by Gunther Prien in *U-47* on the night of October 14 1939. Prien himself was later to go to the bottom with his command in March 1941. HMS *Courageous* had already been caught by *U-29*. This was not to be the only capital ship casualty during the war.

The fall of France gave the U-boat Fleet the advantage of Atlantic operations without having to waste time and fuel transitting the high risk area around Scotland. The Luftwaffe was also now flying intensive Focke-Wulf Condor missions in support of, and occasionally independent of, U-boats operating from the Biscay ports. These missions, by units such as KG 40, could in theory be flown from Altenfjord, Norway, to Bordeaux, France, but unhappily for Dönitz co-ordination was never really good. Although convoys could be spotted and U-boats directed to them, this co-ordination had only a limited success. Indeed, the Condor aircraft used were very often successful themselves as bombers, being able to fly higher than the escorts' limited anti-aircraft fire – until the catapult aircraft merchantmen arrived, then they were only shadowers.

From June to October 1940 came the first 'Happy Time' for the U-boat commanders, when inadequate and untrained escorts gave the boats good hunting amongst badly disciplined merchantmen. However, Axis U-boat skippers, under the tactical control of Dönitz, were encouraged to pour out operational chit-chat by radio. This enabled Allied wireless intelligence's submarine plotting room to keep track of the U-boats and give a precis of data to enable convoys to be diverted and eventually to home anti-submarine hunting groups in on the packs. During this time several Italian submarines were based at the French Atlantic ports, but when Italy surrendered in1943 the U-boats built for the Italian Navy were taken back by the Kriegsmarine.

Although the United States did not enter the war until the end of 1941, the covert assistance of both men and warships did begin to turn the tide against German submarines. With the actual entry of America into the conflict, U-boats now had the US coastal zone to hunt in, this hunt being made more profitable by the US Navy's reluctance to use a convoy system. In January 1942 the second 'Happy Time' began because they were virtually unopposed. In fact, for a few days the US authorities thought that the merchant ship losses were due to mines rather than German torpedoes! For several days, the U-boats were able to operate on the surface, even in daylight, because of the lack of anti-submarine patrols. This was very different from the existence in the Atlantic.

Although Dönitz had always been ordered not to attack any shipping in the neutral zone around the US Eastern Seaboard, he was prepared for US operations and almost immediately after Pearl Harbor sent five large boats into the area off Long Island. These five were later reinforced to give 15 submarines on station to attack unescorted coastal traffic.

At a secret point some 1,000 kilometres west of New York City, Dönitz's 'Milkcows' had established a rendezvous for replenishing the U-boats working the US coast; these now even included the Type VII. These boats were built to a minimum operational size and not really intended for far-oceanic areas. Some of them were later coated with rubber to combat the Asdic location devices of British and Commonwealth warships.

Gradually Allied counter-measures were encountered, especially in the important Halifax to New York areas. With this rather more hostile reception, the BdU (Befeheshaber der Unterseeboote or C-in-C Submarines) ordered his boats south into the warmer climate of the Caribbean. Here the escorts were very few and far between and there were also Vichy French outposts such as Martinique. One particularly bold commander – Korvettenkapitän Hartenstein – took *U-156* inshore at the Dutch island of Aruba. He surfaced in the early morning and prepared to shell the Royal Dutch Shell refinery. However, the main artillery piece, the 8.8 cm general-purpose gun, had a watertight tampon in the muzzle and when the gun was fired, it exploded, severely injuring the gunnery officer. The attack was called off and the injured man was landed at Martinique, but *U-156* reached its home port of Lorient unharmed. If the attack had been successful – and in all probability it would have been because of the proximity to the sea of the finished product tank-farm – then the Allied fuel supply situation in the Caribbean would have been seriously affected – not to mention morale.

However, as the technology of surface and air escorts improved dramatically with radar, the U-boats were forced to move further afield. This move to the Southern Atlantic and the Indian Ocean eventually strained the lines of communications to their limit. The ten tanker U-boats (Type XIV), or 'Milkcows', had been substituted for the surface supply ships which had been successfully rounded up in 1942 by surface and air units with the benefit of the 'Ultra' secret intelligence. Actually the German merchant raiders used for supply were quite a scourge in the early war years, and had been present in the South Atlantic in some numbers. These disguised warships, such as the *Atlantis,* were equipped with two Arado floatplanes and, in this particular case, the old battleship *Schlesien's* 5·9-in guns. Unfortunately for the German Navy, when the Type IXC, *U-126*, was refuelling from the *Atlantis* on a November day in 1941, the Royal Navy's 'County' Class cruiser, HMS *Devonshire,* also arrived on the scene at the same time. Luckily for the Kriegsmarine, the U-boat escaped when it immediately submerged, but the *Atlantis* was sunk. A similar fate overtook the supply ship *Python* which came running in response to the distress calls. This time the *Dorsetshire* appeared on the scene and the RN claimed another victory, but she was chased off by the prompt action of *U-68*'s Kapitän zur See Merten. In the first six months of 1942, 585 merchantmen and escorts had gone to the bottom, but this was not to be enough. The South Atlantic operations around South Africa were blessed with little resistance. Not only were new records for Allied merchant sinkings made but the long-range boats also met on the surface, in daylight, to exchange news – something they had not done since the early days of the mid-Atlantic air gap. Eventually the lack of supplies due to 'Milkcow' sinkings forced these longer duration patrols to be curtailed. The U-boat war was an offensive war and at least 700,000 tons per month needed to be sunk, but at this high peak of action only 513,500 tons per month were doing down.

Now, in late 1942, the U-boats had to contend with the advances of Coastal Command operational readiness in equipment and tactics. In this regard, the Allies were lucky to have the co-operation of the US and RN forces in the Atlantic. One innovation was the Leigh Light, an illuminating searchlight fitted to the wings of long-range aircraft which enabled them to pick out U-boats travelling on the surface, at night, whilst recharging their batteries. Secondly, the new anti-surface vessel (ASV) radar became available to aircraft, a similar set to that already being fitted to the escorts. The combination of these two caused the loss from operations of *U-165, U-502, U-578, U-705* and *U-751* in the Biscay area. The German counter-measures included a radar receiver called the 'Biscay Cross', more anti-aircraft armament and new tactics.

The Biscay Cross was a home-made wooden base with wires strung around it giving a range of about 25 km. Should a signal be received the cross was rapidly taken down below, to be used another day, before the boat crash-dived.

During the last quarter of 1942, over 1,800,000 tons of merchant shipping had been sent to the bottom as a direct result of U-boat operations. However, in the same period, over 40 submarines had been sunk.

The Torch landings in North Africa temporarily left the distant operational areas devoid of escorts and U-boat kills started to mount again, even though German submarines had been diverted to the Straits of Gibraltar area. However, the climax was on its way.

Allied wireless intelligence was enabling the score of kills to mount; for example, the Type XIC submarine *U-158* was detected by several US and British radio-direction finding bases in the Caribbean. *U-158* was chatting to her control when she was caught on the surface, depth-charged and sunk by an aircraft of VP–74, a USN squadron. The U-boat sinkings were slowly mounting, even though they were being successful in new areas. However, these new operational areas were not at all to the liking of Admiral Dönitz who had always advocated a total commerce-raiding war in the North Atlantic.

Even in the later months of 1942 there was still a small mid-Atlantic air gap, because although the RAF's Coastal Command had sufficient medium-range aircraft, such as the Whitley (a converted bomber), the American-built long-range Liberator was only in service with one squadron. It was obvious that carrier-borne aircraft would offer the optimum for total air cover.

During these hectic months of the war at sea, six pre-war Transatlantic liners, including the two *Queens,* had been engaged in high-speed Atlantic crossings ferrying badly needed troops and supplies for that great dream, the Second Front. The Admiralty took great care to route them away from trouble and as they sailed independently of the other merchant convoys, they were rarely sighted by friend or foe.

In early 1943, inspired by the intuitive planning and directions of Dönitz, the Wolf Packs of U-boats in the Atlantic were making spectacular attacks, including sinking seven valuable tankers out of nine which had left Trinidad for Gibraltar. Added to this Wolf Pack tactic, the German decrypting organisations were working at full stretch and they were now able to understand many Allied convoy routing bulletins. However, it is interesting to note that neither side was fully aware of the other's secret intelligence work. In June 1943 the Allies revamped their codes and from then on the Germans were very much in the dark.

Wolf Packs had definite names, such as 'Delphin' (or Dolphin) which attacked the Trinidad–Gibraltar convoy mentioned earlier. Their operational pattern seems to have been a surface reconnaissance line, several hundred kilometres in length, but they were still considered to be one pack.

A major change in the Kriegsmarine organisation occurred in January 1943 when, after a dispute with Hitler, Grand-Admiral Erich Raeder resigned and was replaced by General-Admiral Karl Dönitz, who for a time wore both the hat of C-in-C Kriegsmarine and C-in-C U-boats.

The Battle of the Atlantic, as history has named this bloody fight for survival, reached a climax during the first five months of 1943. The merchantmen sinkings were beginning to decrease and so the Kriegsmarine decided to risk heavier losses to sink more convoys. But in reality, the pressure was now on the U-boat service with additional US Naval forces being made available as Escort and Hunting groups. The US Navy, it should be remembered, had now had time to build up its efficiency after years of isolationism and was now taking an ever-increasing toll of enemy submarines which added to the victories of the Royal Navy and the Royal Canadian Navy. It did, however, take the USN some time to perfect anti-submarine methods, mainly because the Americans preferred to learn their own lessons rather than adopt the practices of the RN, which had been hard won since 1939.

Escort Carriers had now effectively closed the mid-Atlantic air gap and convoys could look forward to almost total air cover during their Atlantic crossing. These carriers were often converted grainships or tankers carrying four or so Swordfish biplane aircraft. An aircraft in the air around a convoy resulted in any shadowing U-boat keeping well down, with the result that when it next surfaced the convoy may well have been out of sight.

In the middle of March 1943 the greatest convoy battle took place, when the Germans decoded an Allied convoy re-routing signal which allowed the 'Sturmer Wolf' Pack to attack two convoys simultaneously when the fast eastbound convoy HX229 overtook the slower SC122. Actually the German Navy was under the impression that they were attacking one large convoy, when in fact they were attacking two smaller ones. When nearly 100 ships had converged the 'Raubgraf' and 'Dranger' packs also attacked, giving the submarine forces 44 units to deploy for a battle which was to last four days. The result was the loss of 21 Allied ships for just one U-boat sunk. It was unfortunate for the Allies that 'Lady Luck' had temporarily deserted them–the Ultra intelligence had a periodic 'blackout' and the convoys were in the mid-Atlantic air gap with no Escort Carrier aircraft able to operate. (Ultra was the code-name for the British decyphering of German radio transmissions.)

Furious convoy battles raged in the Atlantic, especially in the Newfoundland/Greenland/Iceland triangle through March and April. At this time there were 254 operational U-boat units, although only about 70 per cent were operating the North Atlantic convoy belt. More U-boats were to be found training and running trials in the Baltic. However, German morale was being affected by the more determined air offensive against submarines from the escort carriers, such as HMS *Biter* and USS *Bogue*. So much was morale affecting the 'warrior' spirit that U-boat Command found it necessary to remind submarine commanders of their duty–something quite unthought of in 1940.

If the convoy battle of HX229 and SC122 was a defeat, the battle of ONS5 in April was certainly a victory. With two packs totalling 41 U-boats against a depleted surface escort, the last large-scale engagement was won for the loss of 12 merchantmen in exchange for seven U-boats sunk, five severely damaged and 12 boats badly shaken. That sort of ratio could not continue.

Figures show that, of 144 U-boats sunk between September 1942 and May 1943, 118 were lost in the Atlantic theatre; five in September, 13 in October, six in November, six more in December, five in January 1943, 15 in February, 16 in March, 14 in April and no fewer than 38 in May.

Another counter-measure adapted was the Dutch invention of the late 1930s, the *Schnorkel* or underwater breathing apparatus. The effectiveness of this device is shown by the fact that slightly modified snorts, as they are now called, are a part of some contemporary diesel submarines. But without refuelling, these U-boats could not operate far afield and, by June 1944, the last of the two remaining 'Milkcows' had met its doom. U-boats had also been travelling in groups on the surface across Biscay's Black Pit, to attack Coastal Command aircraft! But the aircraft then merely held off and called in a surface escort vessel.

The Germans now deployed the *Zaunkönig* acoustic torpedo which homed on a ship's propeller noises. The British called this development the Gnat, and were not really worried by it after the initial shock of losing several escorts. The Germans actually lost *U-377* and *U-972* to rogue Gnats fired from other boats in a Wolf Pack. It is interesting to note that, during the pre-Invasion time, the Luftwaffe had begun to co-operate more with the U-boat arm–assistance which could have been more beneficial and even decisive earlier in the war.

The Allies had made a significant advance in anti-submarine weapons at this time, including the 'Hedgehog' and 'Squid' systems which throw the charges in front of the attacking vessel. The explosive charge had also been improved from Torpex and replaced by Minol. The latter could apparently crack the pressure hull if the charge exploded within 25 feet of a U-boat.

It is noteworthy that the 'Squid' system is still in use in the world's smaller navies, well over 30 years since it was first introduced.

Although badly mauled, the Kriegsmarine still managed to put 126 submarines into the arena in May. By May 24, Dönitz realised that although convoys were being sighted, attacks were not being pressed home and

U-boat losses were still mounting.

After so many months of bitter and bloody conflict, Dönitz had to report to Hitler that the losses were so heavy that the U-boats must be withdrawn from the Atlantic so that the shortcomings in the volounteer crews and anti-aircraft defences could be remedied.

So now the U-boats were more and more on the defensive. They were not beaten yet, but never again did a quarter's merchant losses figure reach the 300,000 tons mark. The submarine losses, however, reached peaks during the Invasion months of 1944 and again in the last months of the war. So serious now was the lapse in the offensive U-boat war that Hitler had to tell the German people, in January 1944, that it was due to a British device which would soon be counter-measured. Although no one invention was to blame, Allied technology had improved beyond German levels; ingenious solutions to these new devices included using balloons to take silver foil aloft whilst attached to buoys, thus foxing both surface vessel and aircraft radars. Asdic could be deceived by releasing air bubbles on a special 'pig' device which quite successfully deceived the shipborne operators.

When D-Day came, the U-boats operating from Biscay ports numbered about 130, of which possibly a third were unsuitable to prevent a Channel invasion. About 20 units were based at Brest, the nearest U-boat port to the invasion beaches.

The Allies were fortunate in two respects: first, the German High Command was unaware of the actual Invasion route and, secondly, because inter-service co-operation had never been good. At any rate, the Allied anti-submarine patrols were such that very few U-boats made it to the 'Spout' passage area of the Isle of Wight.

The crew of *U-763* managed by accident to do what other crews had perhaps only dreamed of doing, when the boat commanded by Kapitanleutnant Ernst Cordes surfaced in Spithead! Having been attacked near the Channel Islands, he thought, after 36 hours of continued attack, that he had escaped, but when he moved north and ran into a sandbar he surfaced and probably saw the Isle of Wight ferry! Cordes returned to Brest and then left again for Norway, which was safely reached in August 1944. *U-763* survived until she was caught by Soviet aircraft at Königsberg in January 1945.

By August 23 1944 the French U-boat bases had been evacuated and the boats now operated from Norway. *U-953*, a Type VIIC, was the last boat to leave Brest and she actually survived the war to surrender at Trondheim on May 20 1945. The task was, however, now impossible as the Allied offensive might was vastly superior and vital supplies were almost impossible to obtain.

Ironically enough, it was when the U-boat was almost beaten that the Luftwaffe began to increase its reconnaissance support for the Navy by locating convoys in the British coastal rendezvous points. The technique seems to have been the detection of a convoy America-bound and shadowing until its trans-Atlantic bearing could be established, then, notify any U-boats in the area.

There was no doubt in the minds of the senior officers in the Admiralty that the U-boat men, even in those days at the end of the war, were dedicated and disciplined, but they were not necessarily Nazi-indoctrinated. Several well-informed observers were worried in case the Allies lowered their guard and left convoys unprotected and vulnerable, with the escorts in home waters. The Germans did in fact manage to have some of their large, fast 'U-cruisers' operational in 1945, but when the war ended only two of these giants had been at sea for any length of time. One, *U-4706*, a Type XXIII, was actually taken over by the Royal Norwegian Navy and renamed *Knerter* after surrendering at Christiansand. The Type XXI, a streamlined hull ocean-going variant of the XXIII, made more impression but, in the end, they fared badly at the hands of maritime aircraft, such as *U-2524* which was sunk with cannons and rockets off Aarhus.

On May 7 1945 *U-2336* sunk the last merchant ship to be torpedoed, although perhaps the fishing boat which trawled up a wartime German torpedo, which then exploded and sank her, was really the last victim. That was as recently as July 1965! On the other side, *U-320* went to her grave on May 7 west of Bergen–the last U-boat to be sunk by enemy action. It can now be confirmed that no enemy submarine slipped across the Atlantic with Nazi war leaders–or if it did, it surrendered because no U-boat was unaccounted for although several were lost through unknown causes.

ABOUT THE PHOTOGRAPHS

The photographs in this book have been selected with care from the Bundesarchiv, Koblenz (the approximate German equivalent of the US National Archives or the British Public Records Office). Particular attention has been devoted to choosing photographs which will be fresh to the majority of readers, although it is inevitable that one or two may be familiar. Other than this, the author's prime concern has been to choose good-quality photographs which illustrate the type of detail that enthusiasts and modellers require. In certain instances quality has, to a degree, been sacrificed in order to include a particularly interesting photograph. For the most part, however, the quality speaks for itself.

The Bundesarchiv files hold some one million black and white negatives of Wehrmacht and Luftwaffe subjects, including 150,000 on the Kriegsmarine, some 20,000 glass negatives from the inter-war period and several hundred colour photographs. Sheer numbers is one of the problems which makes the compilation of a book such as this difficult. Other difficulties include the fact that, in the vast majority of cases, the negatives have not been printed so the researcher is forced to look through box after box of 35 mm contact strips – some 250 boxes containing an average of over 5,000 pictures each, plus folders containing a further 115,000 contact prints of the Waffen-SS; moreover, cataloguing and indexing the negatives is neither an easy nor a short task, with the result that, at the present time, Luftwaffe and Wehrmacht subjects as well as entirely separate theatres of operations are intermingled in the same files.

There is a simple explanation for this confusion. The Bundesarchiv photographs were taken by war correspondents attached to German military units, and the negatives were originally stored in the Reich Propaganda Ministry in Berlin. Towards the close of World War 2, all the photographs – then numbering some $3\frac{1}{2}$ million – were ordered to be destroyed. One man in the Ministry, a Herr Evers, realised that they should be preserved for posterity and, acting entirely unofficially and on his own initiative, commandeered the first available suitable transport – two refrigerated fish trucks – loaded the negatives into them, and set out for safety. Unfortunately, one of the trucks disappeared en route and, to this day, nobody knows what happened to it. The remainder were captured by the Americans and shipped to Washington, where they remained for 20 years before the majority were returned to the government of West Germany. A large number, however, still reside in Washington. Thus the Bundesarchiv files are incomplete, with infuriating gaps for any researcher. Specifically, they end in the autumn of 1944, after Arnhem, and thus record none of the drama of the closing months of the war.

The photographs are currently housed in a modern office block in Koblenz, overlooking the River Mosel. The priceless negatives are stored in the basement, and there are strict security checks on anyone seeking admission to the Bildarchiv (Photo Archives). Regrettably, and the author has been asked to stress this point, the archives are *only open to bona fide authors and publishers, and prints can only be supplied for reproduction in a book or magazine*. They CANNOT be supplied to private collectors or enthusiasts for personal use, so *please* – don't write to the Bundesarchiv or the publishers of this book asking for copy prints, because they cannot be provided. The well-equipped photo laboratory at the Bundesarchiv is only capable of handling some 80 to 100 prints per day because each is printed individually under strictly controlled conditions – another reason for the fine quality of the photographs but also a contributory factor in the above legislation.

Right *U-103* running on the surface near the French coast. The 8.8 cm deck gun is on the right and the small 'foot' to the conning-tower is the magnetic compass compartment (79aMW/3928/19a).

THE PHOTOGRAPHS

Left Most probably a posed photograph, this shot of a commander at the periscope of a U-boat is exactly how we all imagine a skipper to look. This is Kapitän-Leutnant Schnee of *U-201* (70aMW/3495/3).

Below left Kapitän-Leutnant Kentrat (*U-74*) negotiates a mooring operation in an Occupied French port, whilst the crew secure the fenders and mooring wires (74aMW/3691/10a).

Right The bridge of *U-552* during an Atlantic patrol. The lookouts are wearing the standard submariners' overalls (74aMW/3676/26a).

Below The homecoming; this time to a captured French Atlantic port on the Biscay coast. These occupied harbours saved the dangerous transit north of Scotland (74aMW/3691/4a).

Above Keen concentration on the bridge as Kapitän-Leutnant Kentrat (white cap) cons *U–74* alongside. Note the pennants denoting tonnage sunk during the cruise. (74aMW/3691/9a).

Below The roaring bull, first adopted by Günther Prien, was used by several commanders in the early years and became the emblem of 7 U-Flott (74aMW/3679/5a).

Above The bridge of *U-552* has become a little crowded and the relaxed attitude of lookouts and the presence of a cameraman suggests a homecoming in 'safe' waters (74aMW/3676/28a).

Below The same boat, with a surface escort behind, poses for photographs (74aMW/3679/3a).

Above A submariner coming out of the so-called galley hatch in the after casing (75MW/3712/27a).

Below Two Type VIIC boats alongside on March 25 1942. The outer boat is probably *U-588* (Kapitän-Leutnant Vogel) which was sunk in the Atlantic during July of that year. The other boat is unidentified (75MW/3725/30).

Above Conning-tower detail of a Type VIIC—the French language billboards on shore confirm a Biscayan port (74aMW/3676/23a).

Below Sun-ray lamp treatment for a U-boat's crew. The Kriegsmarine took great care of their élite (75MW/3716/18a).

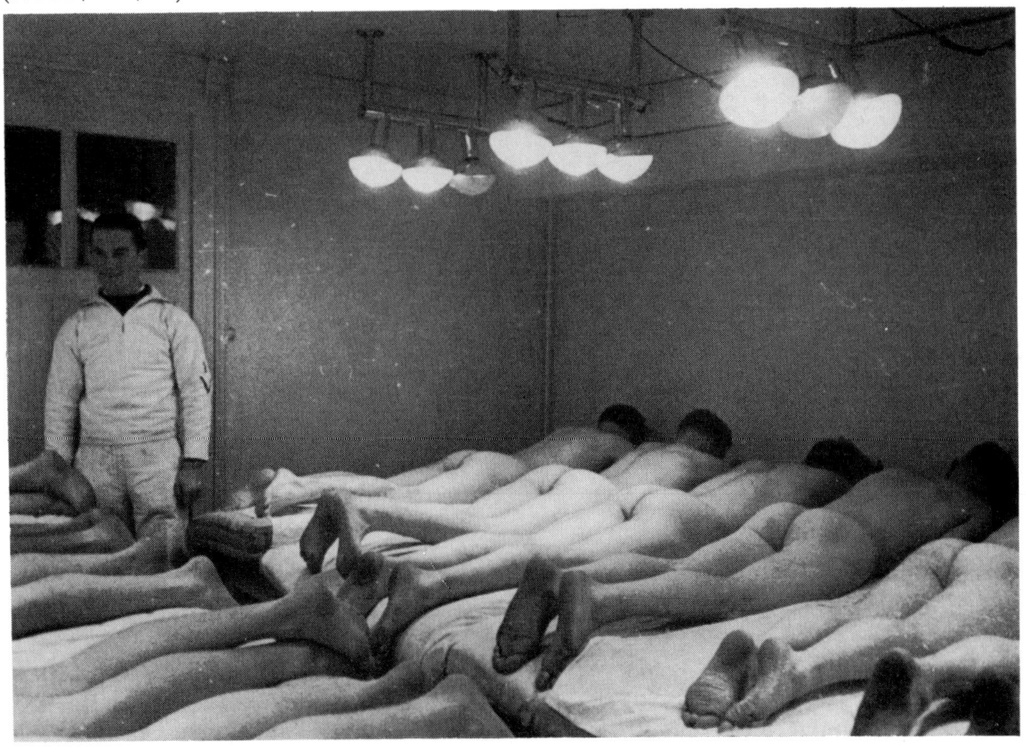

Above left The construction of the U-boat pens at St Nazaire was such that they could only be penetrated by outsized special bombs, but not until 1944/45. In this picture, construction work is still continuing (75MW/3707/14a).

Left Snug as a bug in a rug—two boats in the St Nazaire pens. All the necessary services were laid on for the complete maintenance of submarines. The pens could even be drained down to form dry-docks when necessary (75MW/3707/8).

Above The welcoming home party at Lorient in June 1942 for *U-66* includes some Japanese visitors. Co-operation between the Axis navies was never as close as that between the Allies, but the former did exchange tactics and experience (79aMW/3939/15a).

Right A view from below through the main conning-tower hatch. Note the heavy clamps used to secure the hatch when the boat was submerged (79aMW/3948/18).

Above Although operating in the sunny climes of the South Atlantic, submarine crews could not afford to relax for a moment. The Afrika Korps-style sun helmets are noteworthy (79aMW/3930/27a).

Below The bridge detail of *U-588* shows the radio direction finding loop (left), the attack periscope cluttered with tonnage pennants and the sky periscope to the right (75MW/3725/34).

Above A lighter full of 'fish'! The white rings on the nose of these torpedoes are thought to indicate a particular type (79aMW/3941/23a).

Below The raw material of the U-boats' trade. Torpedoes are transferred by floating crane to a submarine (middle distance) (79aMW/3941/27a).

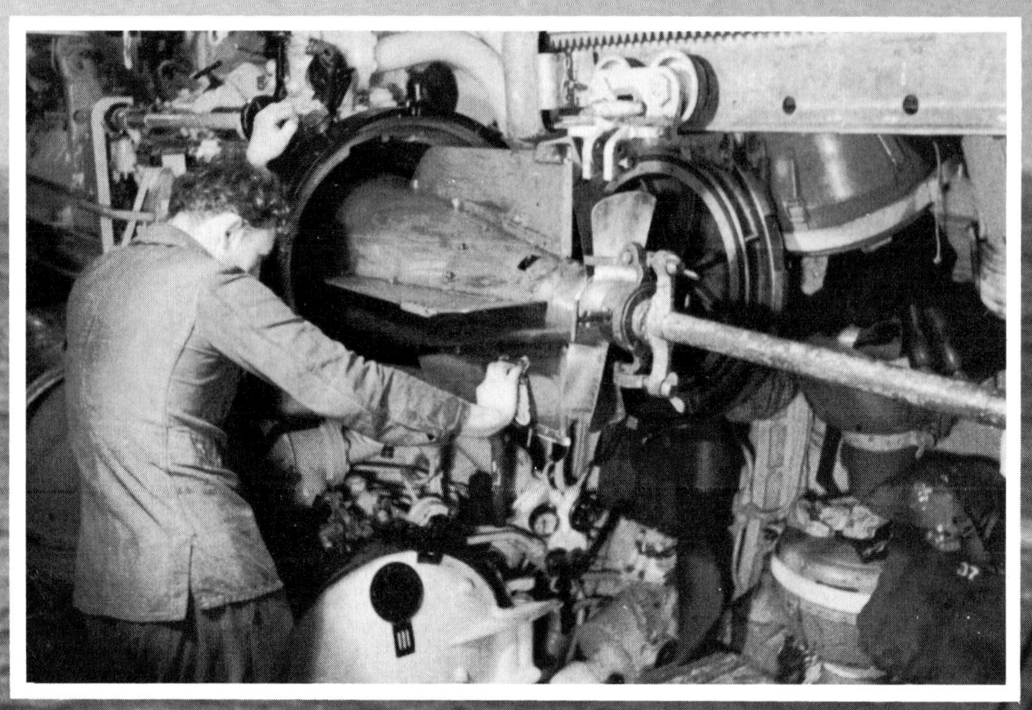

Background photograph The bleak Atlantic scene; the jumping wire and radio aerial leads down to the pulpit at the bow (86MW/4276/28).

Inset above Loading a torpedo tube on a Type IXB U-boat. This class carried 22 torpedoes using four forward and two after tubes for launching the tin-fish (79aMW/3926/15a).

Inset right It was critical when operating in the Atlantic to know the boat's exact position. Here the navigator is computing a fix using data from a sun sight (79aMW/3931/29a).

HOHNER

BERLIN 1930
II
I

Far left The Obermaschinist at the controls in the electric motor control room of *U-103* (79aMW/3927/15a).

Left With his back to a circular pressure door, this crewman whiles away the dull time at sea with a little music. The soft-soled shoes are of interest (79aMW/3929/16a).

Below left The forward torpedo space of *U-103* was also the mess for the torpedo complement who had to live and work in very cramped conditions (79aMW/3926/3a).

Right The container of a famous brand of lubricants has other uses in the workshop of *U-67*. American goods were often used before the United States entered the war and cargoes could sometimes be found floating around in the ocean (80MW/3954/35a).

Below Catching a bit of shut-eye in the forward torpedo compartment. The hammock is slung from the maintenance crane rail. Very often the crew had to give precedence to a torpedo reload in their hammock space (79aMW/3929/18a).

Below right A Christmas tree billows in the breeze as *U-67* arrives home from a successful cruise in time for the seasonal festivities (80MW/3955/26a).

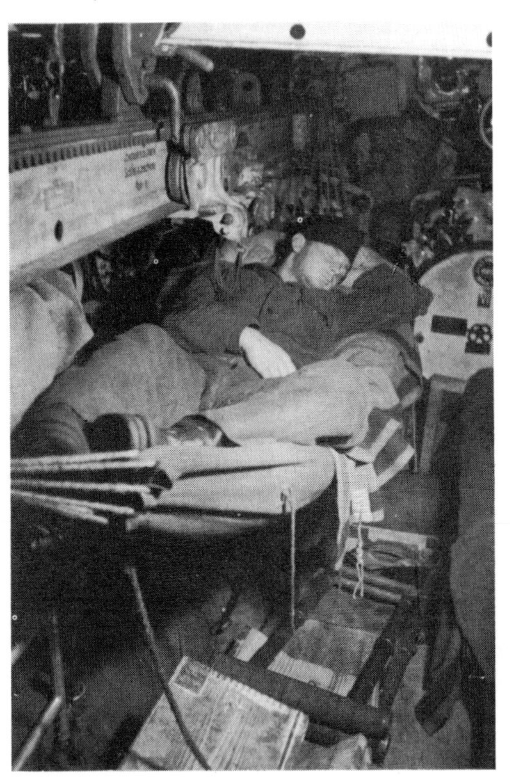

Left *U-37* ties up alongside her depot ship at Lorient in December 1941. Note the rather peculiar inscription 'Westward Ho!' (80MW/3956/18a).

Below Several U-boats were adopted by the towns and cities in Germany and their respective boats displayed their arms on the tower sides. In this case it is *U-37* (80MW/3956/19a).

Right Korvetten-Kapitän Mohr stands below a trophy which is hanging from *U-124's* conning-tower. The life buoy has come from the SS *Sagadahoc* of New York (80MW/3959/32a).

Below right *U-124* returns to Lorient on December 29 1941 with Kapitän-Leutnant Mohr at the con. This boat was lost off Oporto in April 1943 (80MW/3960/10).

S.S. SAGADAHOC
NEW YORK

Left Off on patrol, *U-564* leaves the ex-French naval base of Lorient in November 1941 bound for the Atlantic Convoy routes on one of her first war cruises (80MW/3971/18a).

Right Kapitän-Leutnant Hardegen returns from a successful cruise on February 9 1942. The trophies attached to the jumping wires are of interest (80MW/3981/9).

Below Bedecked with flowers, *U-66* slips into Lorient in June 1942. This boat was sunk off the Cape Verde Islands two years later (79aMW/3939/18a).

Left Christmas goodies on the bridge rim of *U-123* confirm Christmas 1941 as the home-coming time for this Biscayan-based boat (80MW/3994/27).

Below An operational U-boat which did not make it home in time celebrates Christmas the best way it can (80MW/3998/19).

Right Christmas celebrations in a mess aboard *U-67*. Note the Kriegsmarine ensign on the bulkhead (80MW/3954/19a).

Below right Three of *U-103's* officers pose for the camera, whilst a fourth is visible sitting on the tower's wind deflector—a favourite position (79aMW/3928/24a)

Below far right The worn paint work of a U-boat which has returned to port testifies to a winter deployment in the North Atlantic. This boat is *U-66* (80MW/3986/6).

Above left Leaving Lorient for an Atlantic patrol in April 1941 is *U-73*. Note the flowers and buttonholes of a traditional send-off (85MW/4250/7a).

Left In November 1940, Günther Prien, perhaps the most famous U-boat commander, brings *U-47* into harbour. The boat, which sank the *Royal Oak* at Scapa, went to the bottom off Rockall on March 7 1941 (85MW/4242/36a).

Above The navigator takes a sun-sight and another submariner logs the readings. Most U-boat oceanic navigation was by dead reckoning with, hopefully, a sun fix every 24 hours (80MW/3990/37a).

Right A submariner paints up a second Knight's Cross award for *U-123* in February 1942 (80MW/3982/38).

Left The U-boat lookout's standard gear for the Atlantic—sou' wester, raincoat, Zeiss binoculars and tinted glasses (80MW/3951/18).

Below left In early 1942, the deck gun crew of *U-123* are in action against an independently sailing merchantman (81MW/4006/31).

Right Watching the fall of shot as the 8.8 cm gun opens up on its target merchantman. U-boat commanders, in the early war years at least, preferred to use the gun rather than use up their valuable torpedo rounds (81MW/4006/28).

Below The foolish merchantman pays the price of attempting to run the gauntlet of the Atlantic hunting grounds alone. In the early months of World War 2 many ships were lost by the Allies in this way (86MW/4297/37a).

Background photograph The conclusion of a successful war cruise for *U-106*, judging by the kill pennants. She is seen preparing to berth at Lorient on February 22 1942 (81MW/4008/19a).

Inset Foul weather gear was allocated to a particular boat rather than to any individuals. Two types are shown here, the standard oilskins and the heavy weather gear (right) (85MW/4237/13a).

Above left Unfortunately the motif on this submarine has weathered but it is possible to identify it as *U-101* (85MW/4237/31).

Above Daylight watch on the bridge of *U-101*. A constant watch of the horizon was necessary to seek likely targets or spot a threat (85MW/4233/38a).

Left Kapitän-Leutnant Mengersen (*U-101*) on watch during transit to the Atlantic convoy routes (85MW/4229/27a).

Above right A lifeboat of Allied merchant seamen is rescued by a U-boat in mid-Atlantic. The expressions on the men's faces tell all (85MW/4224/19a).

Right Taking sun-sights—probably a Midshipman under instruction during an operational training cruise in the Bay of Biscay. Towards the end, a large number of boats put to sea with inexperienced and badly trained crews (85MW/4222/26a).

Above A routine signal from the U-boat HQ told the crew that Kapitän-Leutnant Schultze was the first U-boat Commanding Officer to be awarded the coveted Knight's Cross and here the Watch Officer is placing a home-made version around his skipper's neck. Note the somewhat anxious lookout! (85MW/4220/19a).

Below When Schultze returned to base on March 9 1941 he was met by General-Admiral Karl Dönitz (C-in-C U-boats), who presented him with his proper award (85MW/4220/30a).

Above In May 1941, 'Edelweissboot', alias *U-124*, at a secret mid-ocean rendezvous to replenish her stores and torpedoes. Note the torpedo stowing gear rigged forward (85MW/4219/19a).

Below left December 29 1941 found *U-124* at sea and undergoing some hurried maintenance to her propulsion gear (80MW/3959/9a).

Below right This crewman from *U-124* shows off the Edelweiss cap emblem, also worn by German alpine troops (85MW/4222/14a).

Left Korvetten-Kapitän Hardegen receives promotion and his Knight's Cross after returning home in *U-123* on February 9 1942 (80MW/3984/6).

Above A happy home-coming for *U-588* as she slides into Lorient. The officer in the white cap is her skipper, Kapitän-Leutnant Vogel, whose white cock emblem is carried on the conning-tower. The date is sometime during February 1942 (80MW/3982/2).

Right Torpedo maintenance carried out in the after torpedo space (80MW/3990/12a).

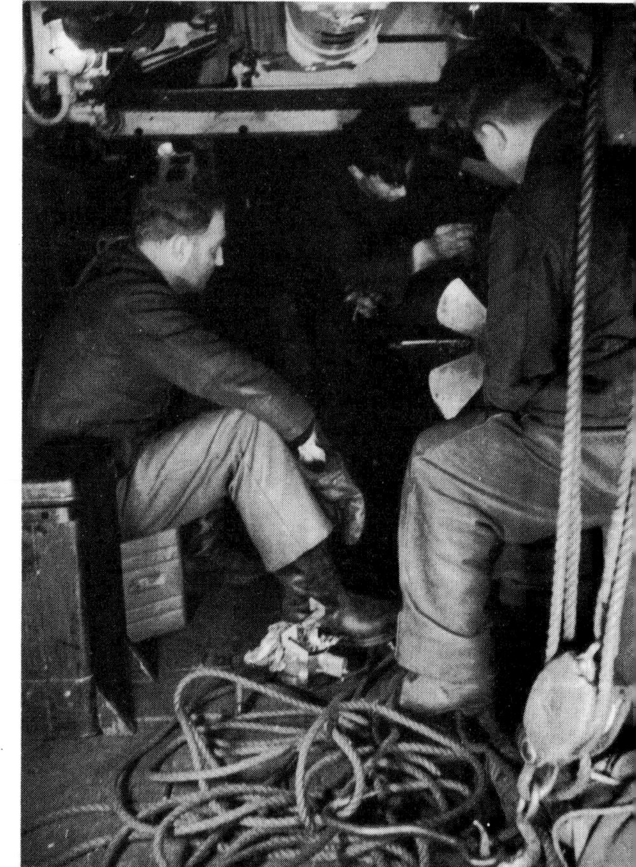

Above left *U-73* arrives at Lorient on February 28 1941 (85MW/4212/27).

Above right Quite where *U-73* collected these bales of cotton is uncertain, but there must have been flotsam and jetsam after each sinking (85MW/4213/12).

Below *U-123*, a Type IXB, coming alongside at Lorient with the smaller Type VIIC (*U-201*) during June 1941 (86MW/4260/37).

Above Two submariners display the emblem of *UA*, the minelaying submarine ordered for Turkey but used by the Kriegsmarine in the Atlantic (85MW/4214/13).

Below The conning-tower detail of *UA* showing the casing around the main gun armament. Korvetten-Kapitän Eckermann watches docking operations on Christmas Eve 1941 (88MW/4366/17a).

Above far left Arriving at Lorient, this Type IXB, probably *U-106*, displays the spray deflector (lower) and wind deflector (upper) on the conning-tower (85MW/4208/19a).

Above left Günter Prien on the tower of *U-47* whilst at sea during November 1940 (85MW/4202/8).

Far left Korvetten-Kapitän Kuppisch with his decorated crew and boat! (85MW/4208/25a).

Left Cramped washing conditions in the forward torpedo space (85MW/4234/28a).

Above Dönitz in conference with his staff. The Battle of the Atlantic received his constant attention even after his promotion to Gross-Admiral in January 1943 (85MW/4201/21a).

Right Kapitän-Leutnant Oesten (*U-106*) is welcomed home by Vice-Admiral Dönitz at Lorient in June 1941. Dönitz tried to reach as many home comings as possible and the morale of the crews showed that his efforts on their behalf were not in vain (86MW/4265/22a).

Left The forward torpedo space was not the best workshop in the world, especially for the maintenance of the complex and cumbersome German torpedoes (85MW/4222/21a).

Below left Dönitz shows off one of his Type VIIB boats (*U-94*) to some high ranking guests in April 1941 (85MW/4210/22).

Right Shark fishing Kriegsmarine-style. Note the watertight tampon in the muzzle of the deck gun (86MW/4286/27).

Below The American tanker *Prairie* refuels a group of U-boats, including *U-107*, just six months before the US entered the war on the Allied side. Prior to Pearl Harbour, many Americans were more pro-German than pro-British and, in fact, several senior officials in major oil companies were naturalised American citizens of German extraction. Indeed, there were four American tankers waiting in the Atlantic to supply the *Bismarck* when she sailed on her last ill-fated cruise (86MW/4285/31).

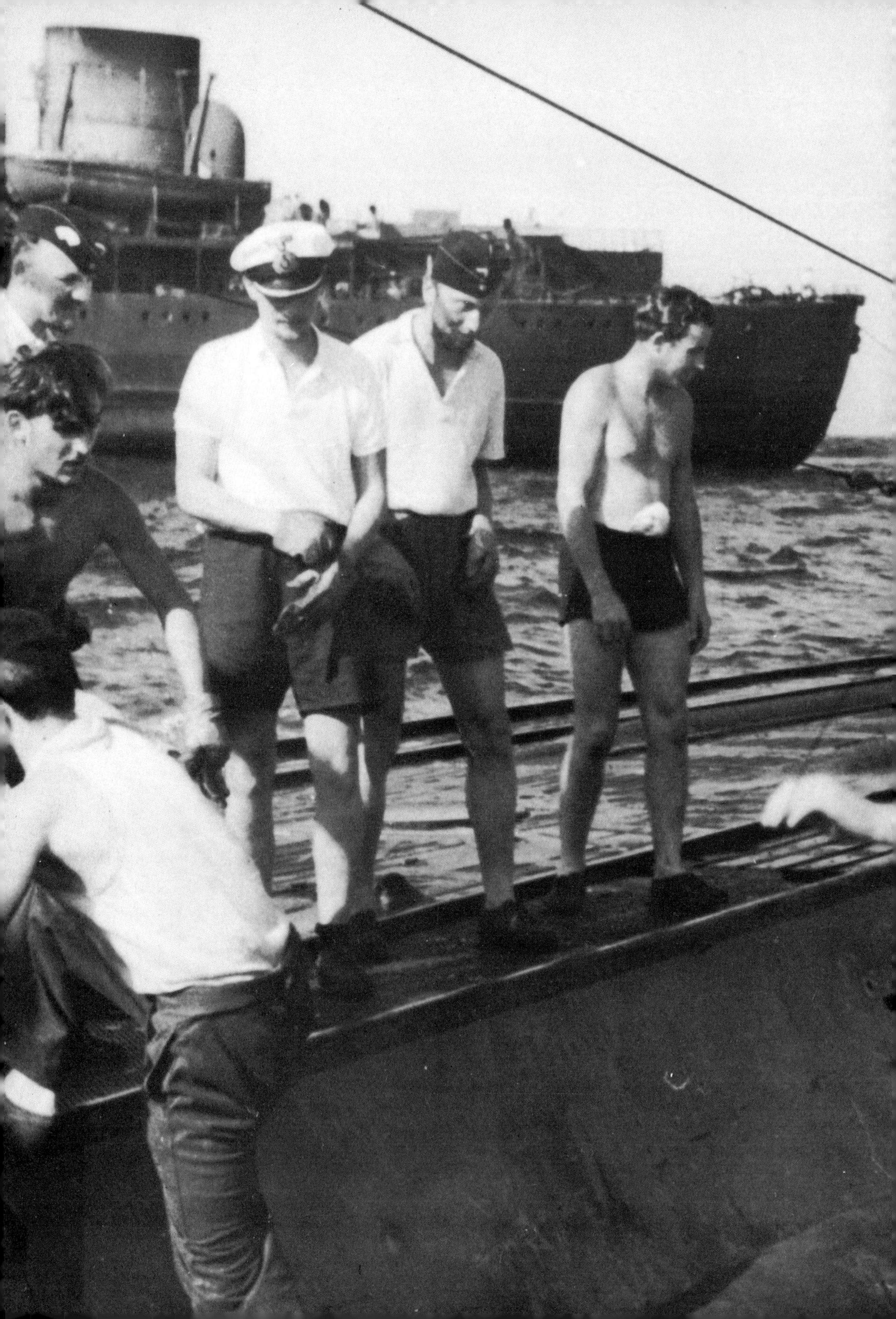

U-69's crew make a hazardous jump across to *U-107* at the Atlantic rendezvous point. Another U-boat runs in to refuel from the tanker (86MW/4285/32).

Above left The pennants strung out from the periscope of *U-107* show just how successful Kapitän-Leutnant Hessler was in the early summer of 1941—a happy time (86MW/4275/10).

Above The four aces badge on *U-107*'s conning-tower with Kapitän-Leutnant Hessler (Dönitz's son-in-law) sitting on the wind deflector (86MW/4272/6a).

Left Before any celebrations the home-coming crew is inspected. The film cameraman and stills photographers tend to detract from the formality of the occasion (86MW/4264/24a).

Right The run into Lorient gave the crew of *U-107* an opportunity to relax on the submarine's casing. Note the mooring wires singled out for berthing (86MW/4275/15).

Above Oesten oversees the berthing at Lorient. The figures on the pennants represent the estimated tonnage of each ship sent to the bottom during that cruise—59,000 tons (86MW/4264/18a).

Below Kapitän-Leutnant Hessler (*U-107*) is decorated and welcomed home to Lorient on July 2 1941 after the most successful war cruise ever (86MW/4272/29a).

Above Oesten tries some fresh June strawberries after arriving home at Lorient. Submarine crews were probably thankful to be home if they had been on a long operation without any fresh fruit or vegetables (86MW/4264/35a).

Below Flowers and German nurses greeted *U-106*'s return, all part of the special facilities granted to U-boat crews (86MW/4264/29a).

Above Thoughts of home? War cruises could be long and lonely voyages but until the very end the crews were volunteers (86MW/4296/16a).

Below A keen eye on the engine room telegraph? Actually this picture does appear to be posed as the gauges are all in a negative position—or could the boat be submerged? (86MW/4296/25a).

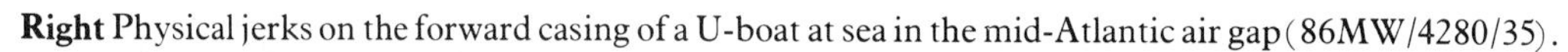

Right Physical jerks on the forward casing of a U-boat at sea in the mid-Atlantic air gap (86MW/4280/35).

Above Stowing a mooring wire on a boat after an Atlantic rendezvous. These were carried out in a stretch of the ocean away from any prying eyes (86MW/4295/23).

Left The sonar apparatus (aft of the bollards) shows up well in this shot of crewmen working on deck (86MW/4295/29).

Above right Submarine maintenance at sea in the Atlantic. Note *U-203*'s deck gun's camouflaged upper surface (88MW/4374/3a).

Right A problem for the boat's engineer as the boat rolls and pitches in the Atlantic swell (88MW/4374/2a).

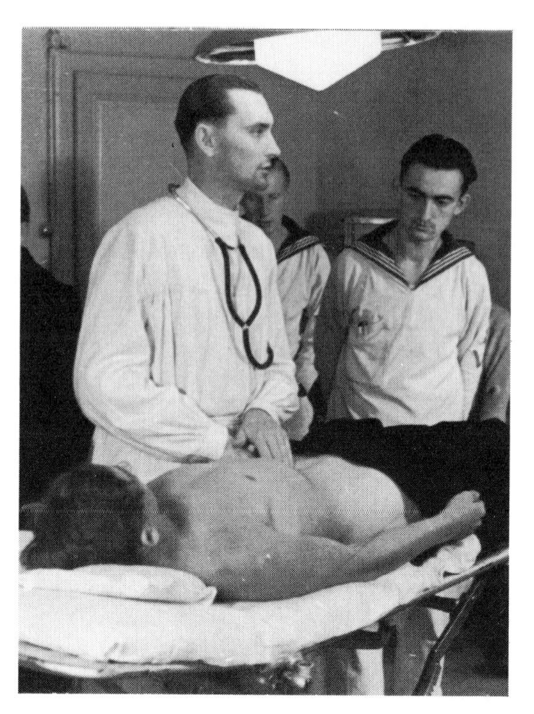

Above far left Far from the prying eyes of Allied aircraft, a U-boat runs on the surface and the crew take a breath of fresh air. The boaters and Afrika Korps sun helmets are somewhat unusual (86MW/4280/19).

Above left The health and physical well-being of the Kriegsmarine's crack submariners was of great importance to the German war effort (88aMW/4393/15a).

Left *U-67*'s pirates seem to be enjoying washing-up aboard their boat after a celebration. Unfortunately we do not know what was the cause of the event (88MW/4370/18a).

Above Arriving at La Pallice in April 1941 is Fregatten-Kapitän Robert Zapp (*U-66*), one of the most respected commanders. Note the sailor to his left is wearing the U-boat clasp under the Iron Cross (86MW/4263/9a).

Right Essential cleaning of the 20 mm Flak gun on the 'wintergarden' tower gun platform aft of the bridge position (90MW/4462/31a).

Left The 8.8 cm deck gun of *U-107* in action against a surface target in the Atlantic in May or June 1941. Ready-use ammunition for this gun was stored in watertight containers sunk into the casing (86MW/4297/17a).

Right *U-105*'s deck gun has been damaged in this photograph taken on May 23 1941 —the cause is unknown but it could be that the gun was fired with the tampon still in place, as happened with *U-156* the next year in the Caribbean (86MW/4262/11a).

Below Interior detail: the electric motor control room of a U-boat. These motors were used to propel the boat when submerged but they needed to have their batteries re-charged regularly. This was usually done at night (88MW/4373/34a).

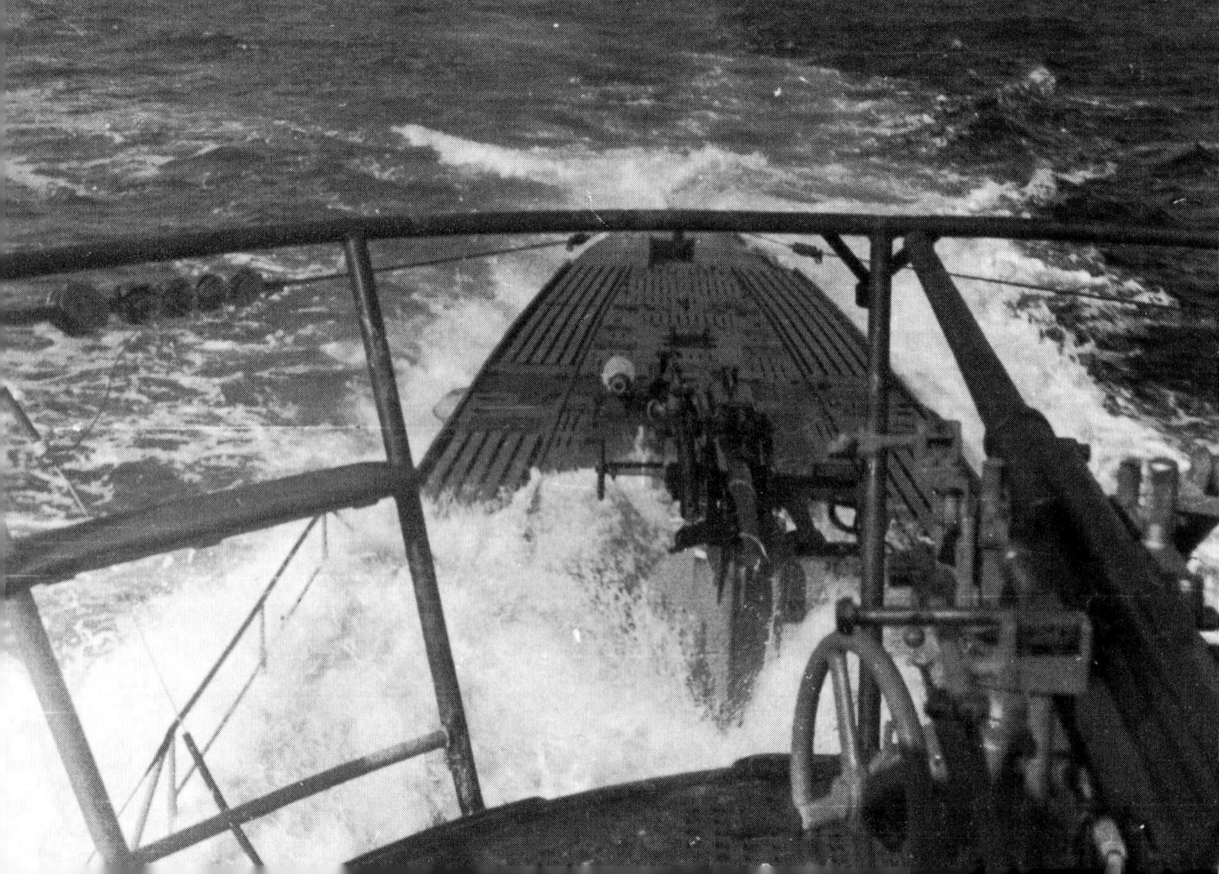

Far left U-boats continuously reported the Atlantic's weather to their Head Quarters. Here a hydrogen-filled weather balloon is being filled prior to launch. Later balloons were used as decoys to fool radar-equipped U-boat hunters (88MW/4373/16a).

Left The polar bear insignia shows up well as *U-108* slides into Lorient on Christmas Day 1941 (88MW/4366/24a).

Below left The view from the rear gun platform looking down towards the stern. The boat is probably a Type IXC (88MW/4369/27).

Right Christmas decorations seem to have been widely displayed as seen here on a Type VIIC's tower (88MW/4366/31a).

Below *U-128* picks a grey day in December 1941 to return to Lorient. Note the paying-off pennant hanging limply from the jack (88MW/4366/3a).

Below right This is either *U-68* or *U-163* arriving at Lorient on July 10 1942 (88aMW/4384/20a).

OVERLEAF

Background photograph *U-107* surfaces to rescue the occupants of a rather crowded life-boat, after sinking their ship in mid-Atlantic (86MW/4287/16a).

Inset Korvetten-Kapitän Johann Mohr keeps watch at sea wearing a suitably marked-up oilskin and *U-124's* 'Edelweiss' badge (88MW/4372/39).

Above left Revictualling in a French Atlantic port. The bottles appear to be mineral water (88aMW/4395/23a).

Above right *U-156* arrives at Lorient on July 7 1942 with some rather strange trophies collected during the cruise. On her next cruise to the southern Atlantic, *U-156* sank the ocean liner *Laconia*. The FuMO-29 rigid radar array can be seen fitted to the tower's leading edge (88aMW/4382/19a).

Below left A practice session with the Dräger Tauchretter or escape apparatus. If a sailor was going to escape from a stricken submarine it was important that he knew exactly how the set functioned (109MW/5429/33a).

Above A hero's welcome for the crew of *U-156*. These welcomes tended to be less glamorous and less glorious as the boats failed to return from operations (88aMW/4382/31a).

Below Japanese sailors visit a U-boat which has just returned from an Atlantic cruise (89aMW/4426/6a).

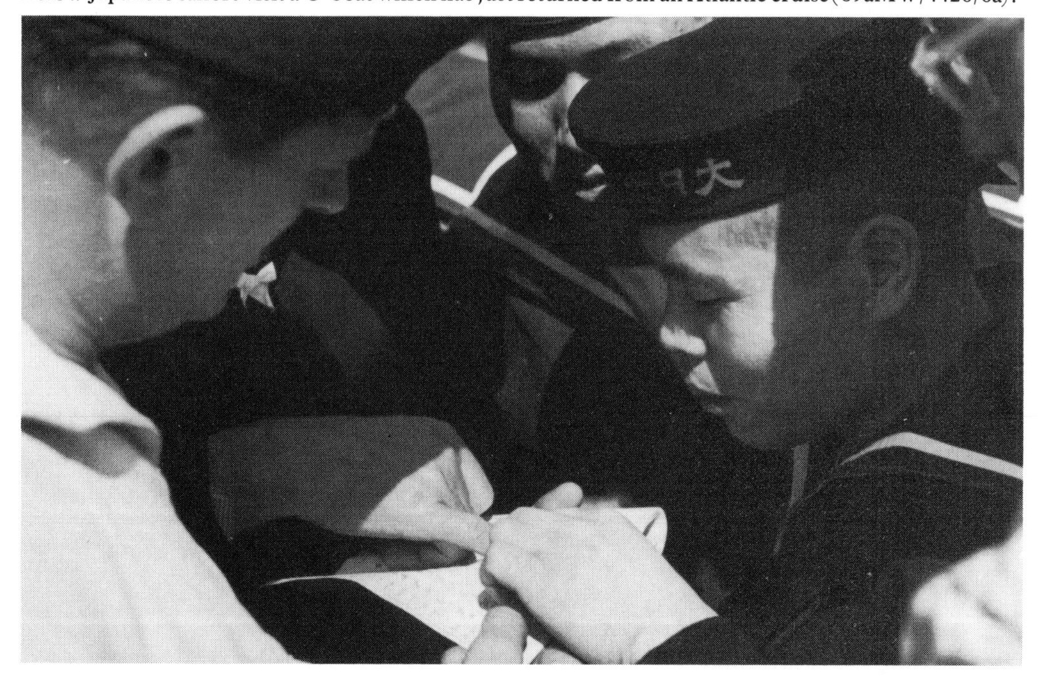

Left *U-132*'s commander sits on the magnetic compass housing to pose for this photograph taken in coastal waters (90MW/4462/17a).

Below Note the signaller on the gun platform bandstand as *U-132* moves into port (90MW/4464/20).

Right A rather battered *U-83* after returning to port with a damaged gun platform, conning-tower and compass housing. The result of a depth charge near miss? (89aMW/4429/2a).

Far right *U-333* has been rammed and damaged whilst creeping through the lines of a convoy. This Type XB boat was actually sunk in July 1944 by the famous HMS *Starling* with the aid of HMS *Loch Killin* (90MW/4457/13).

Below right *U-28* returns to Wilhelmshaven having completed her first war cruise. The date is November 8 1939 (109MW/5429/6a).

Left Surface action! The casing is still awash as the deck gun crew prepare for a shoot. Initial ammunition was brought up from below until the casing lockers were opened (109MW/5438/8).

Above Taking a sun-sight. The block in the foreground is the base of the torpedo-aiming apparatus. The two saucer shapes on the rim are the engine room telegraph repeater (left) and the magnetic steering compass (109MW/5434/22).

Below A second opinion for *U-25*'s skipper? (109MW/5434/18a).

Left Actually before the outbreak of war, wireless aerial maintenance is carried out on *U-29* in a German naval dockyard (109MW/5418/6a).

Below *U-29* stores up for her first war cruise. This makes a useful picture for diorama model making enthusiasts (109MW/5418/10a).

Opposite page Loading torpedoes into an early U-boat at Wilhelmshaven in the winter of 1939. It would be several months before the French bases would fall into German hands and allow the boats to operate from a more forward location (111MW/5536/1 and 3).

This page Preparing for another sortie, this boat has to break up the ice in Wilhelmshaven naval base. Even at sea there was always a risk of icing, not so much in regard to top weight, but the very real danger of the vents becoming blocked and the boat being unable to dive (111MW/5505/24 and 27).

This page Wilhelmshaven on December 19 1939 was a cold place as these two photographs will testify. Note the radio direction finder loop, periscopes and the starboard navigation light (111MW/5509/32 and 33).

Background photograph A Type IXA, *U-38*, seen at Lorient on June 24 1941 after notching up victories in the Atlantic. Note the cupid sitting on a torpedo as depicted in the ship's badge (88aMW/4384/28).

Inset A Type IIB, *U-13* was launched on November 9 1935 and despatched by HMS *Weston* on May 31 1940. The fire-blackened building in the background is of interest and seems to suggest the location is Germany (112MW/5566/28).

Above Wearing naval blue uniform, this Kriegsmarine band pays tribute to fallen comrades (112MW/5568/8a).

Below An impressive gathering of Kriegsmarine officers wearing their formal dress uniform at the solemn occasion of a funeral with full military honours (112MW/5568/20a).

Above The guard of honour fires a final salute (112MW/5568/12a).

Below A depressing sight, but one which became more and more common after the tide of the Battle of the Atlantic turned against the Reich (112MW/5568/22a).

Left The welcoming home committee for this U-boat includes the Befehlshaber der Unterseeboote—Dönitz. *U-129* wears the inscription 'Westward Ho!' (113MW/5613/3).

Below left Dönitz tried to inspire his crews in every possible way and took a personal interest in the progress of each boat (112MW/5564/25).

Right Off to a well-earned leave. The 'kill' pennants are particularly original and include the victim's name and tonnage (113MW/5613/30a).

Below Clearly a propaganda photograph—it would be hard to imagine crewmen returning so soon after they had arrived, as examination of the photo right will show (113MW/5613/28a).

Above Milk comes aboard *U-25* with its shark mouth painted on the tower (109MW/5438/12a).

Below *U-25* showing the zig-zag camouflage and the rather incongruous red and white shark mouth conning-tower art (109MW/5437/29).

Right The crew of *U-124* pose for the official camera on February 21 1942. Several of the crew members wear the oval submariners' clasp signifying several war cruises completed (81MW/4008/8a).

Above A view from *U-50*'s conning-tower as the submarine moves up harbour past some auxiliary minesweepers. Note the fast battle-cruiser *Scharnhorst* in the background (112MW/5566/18).

Left A rather weather-beaten conning-tower (of *U-50*) for these five gallants to pose before. Note the deck gun's breech (112MW/5566/24).

Above right The ship's company of *U-50* gather around the main deck armament after securing alongside their depot ship at Wilhemshaven. The Commanding Officer is instantly recognisable in his white cap (112MW/5567/27).

Right Two submariners, one an officer, wearing their sea-going coats and their newly awarded Iron Crosses (112MW/5567/15a).

Far left Variations on a theme . . . (1) An elephant brooch on the side of this submariner's cap is perhaps a token from a sweetheart, but the wooden torpedo is more difficult to interpret . . . is the wearer a torpedo specialist? (112MW/5567/30a).

Left Variations on a theme . . . (2) This *U-50* crewman prefers the conventional marine anchor and a skull and cross-bones (112MW/5567/22).

Below far left The chamois of *U-67* on the tower with Korvetten-Kapitän Gunther Muller-Stockheim above (88MW/4368/36a).

Below left Having apparently sunk 25 ships (it could be 25 cruises?) the crew of this boat are celebrating with a bottle or two of Armagnac. The bottle is perched on the spray deflector adjacent to the radio aerial intake (90MW/4477/39a).

Right Guarding the important submarine facilities in Occupied France is this naval infantryman and the barrage balloon behind him (111MW/5505/7a).

Below Drydocked in Occupied France, this U-boat gives an impression of its underwater hull not seen when surface running. Note the large rudders (112MW/5581/23).

Above left Camouflage netting is placed over the upper works of this boat as it moves close to the ice. Note the crew's cold weather gear which would indicate either Norway or the Greenland area (506/98/32).

Above right The after casing of an early Type 1A boat at sea in 1939. The crew appear to be re-rigging the radio aerial (109MW/5434/36a).

Below *U-402* leaves La Pallice in October 1942 watched by a party of nurses and other well-wishers. But would it return? (MW/6857/37).

Right Voyaging close to the ice did have the advantage of the occasional piece of fresh meat as this polar bear found to his cost! Incidentally, only the bear's liver is poisonous (506/98/22).

APPENDIX

1. Rank equivalents

To help the reader understand the German Naval ranks used in this account, the following comparison shows the rank the equivalent British Officer would be called by:

Royal Navy	**Kriegsmarine**
Admiral of the Fleet	Gross-Admiral
Admiral	General-Admiral
Vice-Admiral	Vize-Admiral
Rear-Admiral	Konter-Admiral
Commodore	no real equivalent
Captain	Kapitän zur See
Commander	Fregattenkapitän
Lieutenant-Commander	Korvettenkapitän
Lieutenant	Kapitänleutnant
no real equivalent	Oberleutnant zur See
Sub-Lieutenant	Leutnant zur See

The U-boat arm of the Kriegsmarine always held a certain glamour, popularised by propaganda, and this was probably because they were in the front line. Although other branches of the German Navy went into action, for example the S-boats, the majority of Naval operations were coastal convoys. Hitler would never risk the 'big guns' and so perhaps the submarines thought they were contributing in real terms–and indeed they were doing that.

2. Notes on U-boat aces

The World War 2 U-boat aces had the same kind of charisma as the fighter pilots of the Luftwaffe. They were seen by the German populace as having the same warrior instinct.

The aces included Otto Kretschmer of *U-99*, the highest-scoring commander who spent most of his war in a British POW camp, but was the first U-boat commander to receive, on November 26 1941, the award of Oak Leaves with Swords to the Knight's Cross. Gunther Prien of *U-47*, the man who sent the *Royal Oak* to the bottom of Scapa Flow, himself died in March 1941. Admiral Dönitz's own son-in-law, Kapitänleutnant Hessler of *U-107*, carried out the most successful cruise in 1941, the same year as *U-100*, Kapitänleutnant Schephe's command, was lost.

British capital ships fared badly in the hands of the aces, including the *Courageous* sunk by Korvettenkapitän Schuhart of *U-29*; Kapitänleutnant Guggenberger of *U-81* sunk the *Ark Royal* whilst Oberleutnant zur See Freiherr von Thiesenhausen of *U-331* accounted for HMS *Barham.*

They were certainly brave men and most were not the piratical murderers of propaganda as the *Laconia* rescue demonstrated.

3. Notes on U-boat insignia

A study of U-boat photographs, such as the Bundesarchiv collection, results in an appreciation of the espirit de corps of the Kriegsmarine U-boat Command.

Just about every boat had an insignia of one sort or another and these emblems were carried to sea on operational cruises although this was very much against High Command orders.

The emblems seem to fall into four groups. First, the training flotilla tactical signs which were official and carried on the conning tower sides, such as a chequer board. Secondly, the U-Flotilla badges which were usually carried by all boats in a flotilla, in one form or another: for example, the Swordfish of 9 U-Flott, based at Brest. Thirdly, the U-boats which were purchased from collections from the people of towns carried the crest of that town or city. The crest of Köln was carried by *U-208* until her demise off Gibraltar in December 1941. Finally, the individual taste of the boat's commander was often displayed in the emblem, from the skull and crossbones of *U-753* to the Viking ship of *U-83*.

Other titles in the same series

No 1 Panzers in the desert
by Bruce Quarrie

No 2 German bombers over England
by Bryan Philpott

No 3 Waffen-SS in Russia
by Bruce Quarrie

No 4 Fighters defending the Reich
by Bryan Philpott

No 5 Panzers in North-West Europe
by Bruce Quarrie

No 6 German fighters over the Med
by Bryan Philpott

No 7 German paratroops in the Med
by Bruce Quarrie

No 8 German bombers over Russia
by Bryan Philpott

No 9 Panzers in Russia 1941–43
by Bruce Quarrie

No 10 German fighters over England
by Bryan Philpott

No 12 Panzers in Russia 1943–45
by Bruce Quarrie

In preparation

No 13 German bombers over the Med
by Bryan Philpott

No 14 German capital ships
by Paul Beaver

No 15 German mountain troops
by Bruce Quarrie

No 16 German fighters over Russia
by Bryan Philpott

No 17 E-boats and coastal craft
by Paul Beaver

No 18 German maritime aircraft
by Bryan Philpott

No 19 Panzers in the Balkans and Italy
by Bruce Quarrie

No 20 German destroyers and escorts
by Paul Beaver